FARM BUILDINGS

LIFE ON THE FARM

Lynn M. Stone

Rourke Publishing LLC
Vero Beach, Florida 32964

www.rourkepublishing.com

PHOTO CREDITS:
All photos © Lynn M. Stone

EDITORIAL SERVICES:
Pamela Schroeder

Library of Congress Cataloging-in-Publication Data

Stone, Lynn M.
Farm buildings / Lynn M. Stone
p. cm. — (Life on the farm)
Includes bibliographical references and index
ISBN 1-58952-091-2
1. Farm buildings—United States—Juvenile literature. [1. Farm buildings. 2. Farms.] I. Title

NA8201 .S75 2001
631.2'0973—dc21 2001031668

Printed in the USA

TABLE OF CONTENTS

FARM BUILDINGS

Farm buildings protect people, animals, and **crops** from bad weather. They also give farmers a place to **store** things, such as tractors.

Settlers from Europe first built farm buildings in America in the 1500s. These old buildings were made of wood and stone. Later, farmers built with wood, stone, concrete, and metal.

These farm buildings in New York are shelters for crops, animals, and people.

American farms have changed a lot in the last 500 years. Many kinds of farm buildings are no longer used. For example, let's visit a Vermont farm of the 1850s.

That farm would have a house, barn, and **silo**. Modern dairy farms have those buildings, too. But the old Vermont farm might also have a smokehouse and an icehouse. It would have an outhouse and perhaps a **blacksmith** shop. It might have a windmill and many different buildings for animals.

Much of this old Vermont barn was built in the late 1700s.

The smokehouse was where fresh meat was taken. There it was dried, salted, and smoked.

Fresh meat today is packed off the farm. Meat is kept fresh in refrigerators and freezers. Smoked meats are made at **packing plants**.

The farmer's icehouse hid big blocks of ice from the sun. Ice was the only way people had to keep food cold in the 1850s. Refrigerators and electricity ended the need for icehouses.

An old, wooden round barn stands surrounded by newer buildings and steel silos.

The outhouse was an outdoor bathroom. Electricity and modern plumbing ended the need for outhouses.

As farming changed, many other farm buildings disappeared, too. For example, farmers began to raise only one or two kinds of animals. They no longer needed several different buildings for their animals.

Windmills have disappeared from most American farms.

A Holstein cow seems to be reading the sign on the milk house.

This old farm shed may have once stored the rusting truck.

BARNS AND SILOS

The biggest farm buildings are barns. American barns have been built in many shapes and sizes. Some farmers still use barns built more than 100 years ago.

Barns are shelters and homes for cattle, horses, sheep, and pigs. Horse barns are called stables.

Animals live on the ground floor of the barn. They leave and enter the barn through wide doors.

This small building makes a good chicken coop.

Tobacco farms have special barns that help tobacco plants dry. Another plant, hay, is stored in the **lofts** of cattle barns. The loft is like a barn's attic. Hay is stored in blocks called **bales** or as loose straw.

Silos are usually built next to barns. A silo is a big, covered tube. It holds ground-up feed for animals.

A creek becomes a mirror for a 100-year-old dairy barn in Wisconsin.

OTHER BUILDINGS

Most farmers store corn in metal grain bins. Farmers used to store corn in corn cribs.

Milk houses are part of **dairy** barns. A milk house is a room with a large tank. Milk travels in a pipe from the barn to the tank.

The silos (right) on this old six-sided Vermont barn are made of wood.

Sheds are used to store tractors and other farm machines. Sheds also store tools, tires, and wood.

Farms in parts of the Midwest and Northeast have sugarhouses. The **sap** of maple trees is boiled into maple syrup in sugarhouses.

Firewood is stacked outside an old Vermont sugarhouse. The wood will be used to boil maple sap.

Farmers often keep chickens in hen houses, or chicken coops. Chickens lay their eggs in hen houses.

Large, modern hog farms keep their hogs in long, low barns with pens inside. Hogs outside live in little huts called **sties**.

GLOSSARY

bale (BAYL) — a bundle of hay or straw

blacksmith (BLAK smith) — a person who shoes horses

crop (KRAHP) — a field, orchard, or grove of ripening food plants such as corn or apples

dairy (DAYR ee) — having to do with milking cows, milk, and milk products

loft (LOFT) — the upstairs of a barn, often used as a place to store hay

packing plant (PAK ing PLANT) — the factory or building where meat is cut up and packaged

sap (SAP) — the clear, watery liquid made by trees to carry the tree's food to its branches and leaves

silo (SY lo) — a tube-shaped building in which animal food is stored

store (STOR) — to put away or save to use in the future

sties (STYZ) — huts and pens for hogs

INDEX

Further Reading

Kalman, Bobbie D. *In the Barn*. Crabtree, 1996

Websites To Visit

www.billingsfarm.org
www.farmmuseum.org

About The Author

Lynn Stone is the author of more than 400 children's books. He is a talented natural history photographer as well. Lynn, a former teacher, travels worldwide to photograph wildlife in its natural habitat.